A linguistic journey through Western Europe

The linguistic landscape of Western Europe

Oliver Rappitsch

Copyright © 2024 Oliver Rappitsch
All rights reserved

Introduction

Western Europe, a region full of history and culture, forged over centuries of peace and war, also hosts a vibrant linguistic landscape. Even as it may not be as diverse as some other areas on the world, the region still hosts a larger variety of languages.

Not only this, but it is origin to some of the most influential languages on the globe. English, French and Spanish are perfect examples for this.

Historic reasons lead to the widespread usage of those languages around the globe. In Europe itself the modern linguistic tapestry began to take shape with the emergence of the Roman Empire. Latin influenced the local languages and left a lasting impact in all of Europe.

The once widespread Celtic languages survived only in small numbers, while remaining heavily influenced by Latin and the impact of the Roman Empire.

Basque, an isolated language, remains a unique linguistic outliner, offering a glimpse into an ancient language family in a region mostly conquered by Indo-European languages.

Today two major language families can be found in Western Europe: Romance and Germanic. Both are members of the Indo-European language family and use the Latin alphabet.

Romance languages include French, Spanish, Italian, Portuguese and several more. They share similarities in grammar, vocabulary, and phonetics, reflecting their common origin.

Germanic languages include with under English, Dutch, German and some Nordic languages. Formerly they used runes as alphabet, but through the influence of Latin and later on major Romance languages the Germanic languages started to adopt the Latin alphabet. Germanic languages include several dialects and accents influenced by historical and geographical factors.

Celtic languages hold some influence to this day. While Latin's influence on the major languages may be way more obvious and profound, the cultural legacy of Celtic languages endures. Along the shores of Western Europe's coast, one is able to encounter areas where Celtic languages are spoken.

The major Western European languages share common linguistic roots, reflected in grammar and vocabulary, all Western European languages use the Latin alphabet, but often contain special letters, unique sounds and pronunciation patterns. Additionally, eastern Arabic numerals are building the base for the number systems of all European languages, but there are quite some differences when it comes to how those numbers are called and how larger numbers are formed.

To experience the uniqueness of Western Europe's linguistic landscape let us embark on a journey through the countries of Western and Northern Europe.

Romance languages

The Iberian Peninsula

The influence of European languages far exceeds Europe's borders. Through long periods of history both the Spanish and Portuguese empires brought their language to distant continents. They were the first European empires to start with opening colonies on different continents in mass.

Even after their empires fell, the influence of both Portugal and Spain remained, in many ways through the languages they left behind.

Approaching Europe from the west one can already find a series of Portuguese and Spanish speaking islands. These islands contain their own unique dialects, with special words and pronunciation.

On the Canary Islands for example people refer to a bus as guagua instead of the common Spanish autobús. This version can also be found in the territories of some of Spain's former colonies. Moving on to the Iberian Peninsula, and with that to the European continent, one can discover a larger variety of dialects and later on whole new different languages.

Most of them, with the exception of Basque, share a common language family and have influenced each other over a long period of time. Portuguese and Spanish share some vocabulary, but with different pronunciation and sometimes slightly different meaning. Another language on the Iberian Peninsula, Catalan, shares features with Castilian (Spanish) and other romance languages and conveys the same Mediterranean feeling as Spanish.

Basque, as mentioned earlier in the introduction, is a special language. Its isolated state, located in and the near of the Pyrenees, make it a unique language surrounded by Indo-European influence. It is the only surviving language of its language family and offers a glimpse into the past.

The two largest languages of the Iberian Peninsula Portuguese and Spanish share many similarities, visible through many cognates and reasoned in their common heritage not only as Romance languages, but also as tongues of the branch of West Iberian languages.

Entering the European mainland through Portugal the first language we encounter is Portuguese. A world language with around 260 million speakers worldwide, Portuguese evolved from Galician-Portuguese, a common language in the medieval period. Galician-Portuguese is also the ancestors of Galician, which is spoken to the north of Portugal and shares even more similarities with Portuguese than Spanish does. The Portuguese language heavily relies on vowels and contains some special letters. Additionally portions of its vocabulary come from Arabic, due to the long period of Arabic rule on the Iberian Peninsula.

As evident in its traditional music genres like Fado and Samba, Portuese is especially known for its musicality and rhythm.

Nowadays much of the influence Portuguese holds comes from Brazil where most of its speakers live. Through Brazil Portuese is one of the official languages of the BRICS and several South American institutions and through Portugal it is one of the 24 languages of the European Union.

Following the Tejo river until it becomes the Tajo we enter the kingdom of Spain.

Spanish is one of the most widely spoken languages worldwide, with more than 500 million native speakers and around 76 million speakers in Europe. Additionally, it is one of the 6 official languages of the United Nations and according to the Instituto Cervantes it is the third most used language on the internet after English and Chinese. Spanish has a unique phonetic system with letter pairs like ll, rr, gu, qu and ch and the additional letter of ñ.

Including ñ, the Spanish alphabet has 27 letters, 26 of which come directly from the Latin alphabet. K and W are only used in words and names coming from foreign languages for example kilo, kiwi and folklore. Spanish is characterized by its phonetic system, including the use of the seseo and yeísmo phenomena.

Just as with Portuguese, the vocab of Spanish was influenced by Latin and Arabic, which can be seen to this day. Due to the languages global reach, there are many dialects around the world and Spain itself is home to quite a few of them.

Bordering the influence of Castilian is Catalan, spoken in Catalonia, the Balearic Islands, the Valencian Community and Andorra where it is an official language. It has 8 million speakers.

Catalan shares similarities with other Romance language, particularly Spanish. The language is strongly engraved in the cultural identity of its region and plays an important role in discussions about society and politics. Other than the languages named prior it is not part of the Western Iberian branch, but rather a member of the Occitano-Romance branch, like Occitan.

In general, the Iberian Peninsula is home to several related languages, with common vocabulary, but nevertheless unique characters visibly through special letters and different phonetic systems. Travelling through this region one can encounter this variety of dialects and cultures.

In the north near the Pyrenees mountains Basque or Euskera remains a unique linguistic phenomenon. It's an isolated language, with ancient roots that could date back to the Palaeolithic era. It has a complex grammar and has several dialects even though its small geographic spread.

Over time Basque borrowed around 40% of its vocabulary from Romance languages and uses the Latin alphabet. Due to its old age, it uses words from all kinds of language families that have been spoken in Europe since prehistory. With that it offers a glimpse into the history of the Iberian Peninsula and Europe as a whole.

As with Catalan the language is deeply intertwined with the culture of its region, still it remains endangered. The influence of the surrounding Indo-European languages holding strong on it and its speakers.

France + Italy

Heading further to the north we cross the border to France. French is France's most spoken language, but there are more. Right in the south one can encounter Occitan, like French a member of the Gallo-Romance branch of the Romance language family.

As mentioned in the part about Catalan, Occitan is a member of the Occitano-Romance branch, making it a member of the Romance language family. With estimated 1,5 million speakers its is an endangered language.

Occitan has 23 official letters, with the other 3 (k, w, y) only in use in foreign words such as whisky. This mirrors the approach of Spanish. Occitan doesn't have an official status inside of France, neither does it have a single written standard form. As there are several competing norms of writing the language, most efforts to standardize the language are slowed down or hindered by the rapid decline of speakers and by the significant differences between different dialects of Occitan

From the south of France, we move on to the northwest to the peninsula of Brittany where the Celtic language of Breton is spoken by around 200 000 people.

Breton is recognised as a minority language in Brittany and is regulated by the Ofis Publik ar Brezhoneg. As with Occitan it is an endangered language and faced decline in the 20th century. This decline had been especially fast as the number of Breton speakers declined from more than a million to the 200 thousand that are estimated today.

Breton is a Southwestern Brittonic language, just like Cornish in Cornwall, of the Celtic language group. Even though it is the only Celtic language still widely spoken on the European mainland, it is not a member of the continental branch, instead Breton is part of the insular branch. All surviving Celtic languages are from this branch.

Due to its long-standing connection with France, it is estimated that around 40% of its core vocabulary is loaned from French. Despite the dominance of French, Breton remains an integral part of the Brittonic culture and shares much of its vocabulary with Cornish and Welsh.

Talking about France one must talk about French. Often stated as one of the most beautiful languages to listen to, it has a unique pronunciation outstanding between its related neighbours.

With estimated 320 million speakers and official state on several continents French is one of the UN's official languages just like Spanish. It has been a language of diplomacy, culture and arts for centuries.

French has a large variety of regional variations, including Parisian French, Provençal and several others outside of the European mainland such as Québécois.

The language evolved from Vulgar Latin and was influenced by Germanic invasions and the Carolingian Empire. Even though its different pronunciation it shares many similarities with other Romance languages. Many words can be directly translated from other Romance languages and put in use with some simple changes in tone and pronunciation.

France's global influence stems for one from centuries of influence in art and science and the general prestige of the language. For the other colonialism forced large parts of the world under French – and Belgian – control. In many of those areas French is spoken there to this day.

Besides the languages d'oil – French and its closest relatives – and the language d'oc, which we already discussed under the name of Occitan there is another member of the Gallo-Romance languages spoken within France. Its name is Francoprovençal includes several dialects and has around 150 000 speakers in France, Italy and Switzerland.

It was formerly spoken in the duchy of savoy, but now Francoprovençal is mainly spoken in the Aosta Valley in Italy, where all age ranges use it. Due to the fact that outside the Aosta Valley there are mostly to only older speakers of the language and the reduction of speakers of the years it is rated as a definitely endangered language.

In France it is recognised as a minority language in Auvergne-Rhône-Alpes and Bourgogne-Franche-Comté, while in Italy it has a protected status in the Aosta Valley.

Staying in France a bit longer we go to Corsica where one can encounter Corsican. A language spoken by around 150

000 people, just like Francoprovençal it is considered as a definitely endangered language. There is some debate if it should be counted as its own language or a dialect Italo-Dalmatian, but in every case, it is part of the Romance language family with strong ties to the Italian language.

Going further down towards Italy to the island of Sardinia we encounter the Sardinian language. Many Sards are very proud of the Sardinian language. With more than 1 million native speakers it is still considered a definitely endangered language but is in a way better condition than many of the other languages in the same category.

Together with Italian it is considered as one of the languages closest to Latin. The impact of Latin can be strongly felt in the language, just as the Sardian language's own impact can be felt on the island. Italy is also the country where the impact of the Roman Empire can be felt the strongest.

Moving down to Italy's largest island, we encounter the Sicilian language on Sicily and its surrounding islands. As most of the languages named prior it is a member of the Romance language family.

It has limited recognition in Sicily where most of its speakers live, nevertheless Sicilian is known to be taught in schools.

Leaving Italy to the south we stop by the island nation of Malta. Its second official language next to English is Maltese. A very interesting language telling the story of the island even more than its architecture does. Maltese is a Semitic and Afroasiatic language, making it the only official language of the European Union with this background.

But as said all Western European languages are written with the Latin alphabet and so is Maltese, making it outstanding among the Semitic language, as being the only one to be solely written in the Latin alphabet.

Maltese evolved independently from Arabic and went through a gradual process of Latinisation. Over a time, it borrowed more than half of its vocabulary from Romance languages mostly Italian and the prior named Sicilian. Later in the time of French and English rule over the island it adopted some words from English and French too.

The Maltese alphabet is of Latin origin, called the Maltese abjad and includes several special letters, which can be found prominently in place names all around the island.

Moving back to the mainland of Italy, in the north around the city of Naples and most of southern Italy there is the Neapolitan language. A law of the Region of Campania stated in 2008 that the language should be protected.

Italy's main language and an important part of Italian culture is Italian. As we move further through Italy, we experience this beautiful language. Native to Italy, San Marino, the Vatican City and the Swiss Confederation, Italian has around 63 million native speakers, additionally it is also an official language in Corsica, although it has lost in status and prestige there since the time of the Italian annexation of Corsica under Benito Mussolini.

Italian has a 7-vowel sound system, meaning that additional to the five vowels, e and o both have mid-low and mid-high sounds. Most Italian words end with vowels and the language has a contrast between gemination, long consonants and short consonants.

Italian is one of the nearest languages to Latin and can be considered the modern version of Latin with origin in poetic and literary works. The progress of the Italian language was slow and long starting in the 5th century with the fall of the

Western Roman Empire. Predating Italy's unification Italy was slowly made an official language for all Italian states. In the middle ages, through the influence of Petrarca and Dante Alighieri, the Tuscan dialects was established as the most prominent literary language in all of Italy.

In the Renaissance era Pietro Bembo was influential in developing the Italian language. From the Tuscan dialect as literary medium he codified the language for modern usage. Later during the times of Italian unification Alesandro Manzoni helped create linguistic unity throughout Italy and set the basis for the modern Italian language.

Italian influence around the world doesn't stem that much from colonialism as with many other major European languages. In the case of Italian, it was rather migration to distant places that made Italian a worldwide used language, but in a less official and more private way. In Argentina and Venezuela for example Italian is the second most spoken language after Spanish and in the US, there is a significant Italian minority, especially in New York.

In the north of Italy and the south of Switzerland, bordering German speaking areas, is Rhaeto-Romance a purported subfamily of the Romance languages. Three languages would belong to it if it were genuine: Romansh, Ladin and Friulian. Friulian and Ladin are located in Italy, while Romansh can be found in Switzerland.

Ladin and Romansh originate from Latin spoken by Romen soldiers when they conquered Raetia. Friulian is often called Eastern Ladin, but through the influence of surrounding languages it has diverged from Ladin. Ladin itself is spoken in Ladinia, a region mostly located in Trentino-Alto Adige, but also in Veneto.

Romansh is one of the early languages that replaced the Celtic languages spoken in this area. Some Romansh words come from those early Celtic languages, but the by far largest influence in recent times has come from German, which borders all Romansh speaking regions. Romansh is one of Switzerland's four official languages.

After discussing Romansh there is only one official language of Switzerland left: German. In the course of the next chapter, we will be headed further north. As we do this we will delve deep into Germanic languages and their dialects.

Germanic languages

German + Benelux

Delving into the world of Germanic languages we start with the one which's name is already included in the word Germanic: German. By doing so we first enter the area where the Celtic languages originally came from.

As mentioned prior, German is one of Switzerland's four official languages, it is also an official language in Germany, Austria, Belgium, Luxembourg and South Tyrol in Italy and is recognised as a national language in Namibia. The connection to Namibia stems from old colonial times, there the only German daily newspaper outside the European German speaking community is located. It is called the Allgemeine Zeitung and was founded in 1916, making it the oldest daily newspaper in Namibia. In other parts of the world German was and is spoken because of migration of German workers to those places in the late 19th and early 20th centuries.

In Europe German is spoken from South Tyrol in Italy to Northern Germany. Claiming a large part of Central Europe there are also sizeable amounts of German speaking communities outside the prior named countries.

Especially in the western Balkan German is often learned as a second or third language. Other tourist areas, especially some islands e.g. Rhodos or Mallorca, host many German speaking staff so to offer accommodations for tourists from Germany, Austria and Switzerland. While the influence in the Balkan's not only stems from tourism, but also from historical ties and economic cooperations, many other places are only home to a large number of speakers of German because of tourism.

German is the most spoken first language in the European Union and the second most spoken after English if counting all speakers. Globally German is in the top 15 most spoken languages. It has roughly the same amount of L1 and L2 speakers, 100 million each and has had a great impact on science and philosophy. Currently, its influence rather stems from Germany's strong economy and the large number of speakers in Europe.

In contrary to many other languages, German speakers generally don't show as much pride in their language as other's do. Generally speaking, most young speakers of German are capable of conversing in one to two other languages and will do so if its more convenient. This lack of pride can also be seen when looking at the European Union. Even though German is one of its three working languages, most business is done in English and French. From a German perspective this is just natural and makes the most sense and so I won't clarify this any further.

German uses the Latin alphabet, but also includes four special letters: the three Umlauts ä, ü and ö and the sharp-S: ß. The Umlauts are basically a combination of a vowel and e, pronounced fast enough so that they melt into one letter. The ß is basically an ss, with most of the features of an s, which means that it doesn't speed up the pronunciation of the following vowel.

Those special letters aren't included in the alphabet itself but are added to it together with their respective categories. German also uses many diphthongs like eu, au, ei and ai. Larger versions of the German alphabet normally also include at least three of them.

German grammar is known for its complexity, with three genders and four cases and its pronunciation can be seen as harsh. The part about the grammar holds some truth, but the latter statement about the pronunciation is rather rooted in social media trends and an historical rivalry between the German speaking world and other major countries and their languages. The First and Second World War also played their role in forging a slightly estranged picture of German in the minds of some people and nationalities.

Apart from diaspora communities around the world German can be divided in three main dialect groups. Upper and Central German can both be counted to the broader term of High German, while Low German is different enough from High German to be considered its own German.

Our journey first brings us to the area of High German, which contains a large variety of dialects. Reaching from Austrian and Swiss Upper German dialects to High Franconian ones, High German also includes all Central German dialects.

Central and Upper German build the foundation of Standard High German, which is the official standard of speaking and writing German.

A group of varieties of the Moselle Franconian dialect spoken in Luxembourg have been standardized and institutionalized. Now they are considered a separate language, called Luxembourgish. Due to Luxembourgish being based mostly on the Moselle Franconian dialect, it is hard for German speakers of other dialects to understand speakers of Luxembourgish. Additionally, there is a large number of French loanwords in Luxembourgish, giving the language a special flair.

Back to Germany it is important to note that even though the north of the German speaking area holds a unique dialect in form of Low German or Plattdeutsch, Northern Germany is also considered the area where the purest Standard German can be found. There is nearly no influence of the dialect on conversation in everyday life, building a strong contrast to the south where dialects strongly shape the way people speak.

Still there are some expressions that are mostly used in Northern Germany and influence of Low German can still be found in differences in pronunciation, but these are often overshadowed by general regional differences.

To the west of Germany and the areas where Low German is spoken are the Netherlands. There we can find Dutch and the Frisian languages.

Starting with Dutch, the most spoken language of the Netherlands and Flanders in Belgium, we encounter the third most spoken Germanic language. It is spoken by around 25 million people, being native to the Netherlands, Belgium and Suriname. Afrikaans, a partly mutually intelligible daughter language of Dutch is spoken by more than 15 million people in South Africa and Namibia.

The Netherlands had one of the strongest colonial empires at a time, which can still partly be seen through the usage of its language around the globe, although in a different way than with other languages, as the Netherlands didn't lay as much importance on spreading its language and culture as other countries did.

Dutch doesn't use Germanic umlauts and is one of the closest relatives of both English and German. The Dutch vocab is mostly based on Germanic, but there are also some Romance loans. The amount slightly exceeds the one of Romance loans in German but is by far less than in English.

Moving on to the Frisian languages, which are spoken in the Netherlands and Germany, we again encounter a group of smaller languages. Frisian and English are both part of the

Anglo-Frisian language group but are not mutually intelligible.

In general, the three subdivisions of Frisian languages, West, East and North Frisian are not mutually intelligible among each other too, which is owed to language contact with neighbouring languages and independent linguistic innovations. Inside the three subdivisions there are several dialects complicating understanding between the different groups of speakers. Most speakers of Frisian languages live in the Netherlands; therefore, their pronunciation and use of grammar tends to be influenced by Dutch.

The Frisian languages are considered endangered, especially as depending on their location, German or Dutch are simply more practicable for everyday life.

Northern Germanic languages

Bordering Germany, the German speaking area and the North Frisian languages is Denmark with Danish as its most spoken language. Together with Norwegian and Swedish Danish builds a trio of related and to some part mutually intelligible languages – naturally depending on the accent of the speaker. But there are still some notable differences between all three – or four – of them.

Starting with Danish, a language spoken by around six million people, most of whom live in Denmark, we continue our journey on to the Jutland peninsular and the isles ruled by the Danish crown.

Danish has an unusual sound system, consisting of a large inventory of vowels. Depending on the analysis there are up to 27 vowel phonemes. 13 of these are long vowels, 12 short vowels and two are central vowels. Some analyses don't count these last two unstressed vowels, but even if these are excluded, Danish still consists of 25 full vowels. This is a very large number among the world's languages. Like Norwegian it uses the letter ø in writing.

In comparison to the vowels there is only a very small number of consonants in Danish, as the language only distinguishes between 17 consonant phonemes.

Danish is the most German influenced of the Scandinavian languages and has a far more complex grammar than its northern neighbours. For example, there are three different types of regular plurals.

To the west and north of Jutland is the Scandinavian peninsular. Starting in the west of Denmark by crossing the Øresund via the Öresund bridge connecting Copenhagen with Malmö, we enter Sweden. There Swedish is spoken by 10 million people in Sweden and Finland.

Several centuries ago, the Swedish Empire held control over Finland and has left a lasting impact there partly present through Swedish, which is spoken in many areas of Finland to this day. Historically Swedish has also been spoken in Estonia, because of the prior mentioned Empire, but the current status of Estonian Swedish is almost extinct.

The Swedish alphabet has 29 letters, including the 26 letters of the Latin alphabet and the letters å, ö and ä. Å is also found in other Nordic languages, while ö and ä and can be found in German. Some foreign words or names in Swedish retained their original letters. Therefore, the German ü can be found in some words, but is treated as a variant of y instead of being counted as a separate letter.

Swedish is the hardest to understand for speakers of the other two languages and has had less influence from German and English than its neighbours.

Moving on to Norway we discover its charming language. Norwegian is spoken by around 4 million people and is divided in two official written forms: Bokmål and Nynorsk.

Bokmål is based on Dano-Norwegian, which was the elite language in Norway in the 16th and 17th century. Nynorsk on the other hand is based on the written forms of several dialects. Bokmål is the more widespread written form, but Nynorsk aligns more to certain dialects, making it easier to understand for some. Norwegian has been influenced by English, making it easy to learn for English speakers.

In general, Norwegian has an easier grammar than Danish and is best understandable by speakers of both Danish and Norwegian. This effect doesn't always go both ways, especially not with Swedish, but the likelihood to understand something is still way higher than with other more distant languages.

In the cold north of Norway, a small percentage of Norwegians speak one of the Sami languages, even though the small number of speakers, the grouping is still one of Norway's official languages.

The Sami languages are a group of Uralic languages and therefore not mutually intelligible with Norwegian. Depending on how they are counted there are ten or more Sami languages. Most of these are at least partly spoken in Norway, more speakers can be found in Sweden, Finland and the Kola peninsula in Russia.

Leaving mainland Europe, we move on to the Faroe Islands, which are part of Denmark. There the Faroese language is spoken by most of the population living on the islands.

Its closest relative is Icelandic, but they are not mutually intelligible in speech. The written forms resemble each other more, which is largely owed to Faroese's etymological orthography. There are many dialects of Faroese, but there is little evidence that any of them has developed prestige status.

Faroese has 29 letters, all are dived from the Latin script, but with some additions. Some consonants are not present in the language, for example there is no w and no x, instead Faroese contains letters such as ø, ð, æ and some more.

From the Faroe islands we go north. There, part of the Mid-Atlantic Rim, is Iceland. As mentioned, its related to Faroese, but is not mutually intelligible. Icelandic is more distinct from German and English, the most widely spoken Germanic languages.

Even though the Faroese and Icelandic written forms resemble each other, Icelandic still has some special letters apart from those Faroese has and some others are written in a different way, for example it contains the letter þ and x, but just as Faroese doesn't make use of the letter w.

The British Isles

Going back southward, we meet the British Isles, where in the north of Scotland Scottish Gaelic is spoken. With it we enter the world of Celtic languages, which we already partly covered in the part about Breton.

As already said, all remaining Celtic languages are from the insular branch and are therefore related to each other in one way or another. Together with Irish and Manx, it developed from Old Irish and became a separate language sometime during the Middle Ages. Formerly, Gaelic was spoken in most of Scotland, while now it is mostly spoken in the north of Scotland and the Outer Hebrides in particular.

There is an ongoing language revival with the number of young speakers staying relatively stable.

Right now, it doesn't have official standing in Scotland, but the Scottish parliament is considering a bill to give it official status in Scotland. Outside the UK, Gaelic is also spoken in Nova Scotia by a small percentage of the population.

Staying in Scotland, we move to the east to encounter Scots; due to its name, it can easily be misunderstood to be the same language as Scottish Gaelic. In reality they are totally different from each other. For one, Scots is a Germanic and not a Celtic language. It is therefore rather connected to English than to Gaelic or any of the other Celtic languages.

Scots has more than a million speakers, most of whom can be found in Scotland and Northern Ireland. It shares similarities with English, as both languages are from the Angelic branch of Germanic languages, and it is an official language in Scotland and recognised as a minority language in Northern Ireland and Ireland. The status of Scots is not without discussion, with some debating it to be a variety of English.

Moving on to Ireland, we encounter Irish. It is the first official language of Ireland but used by fewer people than English. Irish has several hundred thousand speakers, with its usage being promoted by the Irish government; nevertheless, a large number of speakers only use it within the educational system.

In general, even with the support of the government, Irish has to fight strongly against the overwhelming strength of English.

Historically it used the Ogham writing system, an interesting alphabet with each letter being represented by combinations of one to five straight lines. Nowadays, as the rest of western Europe, the Irish use the Latin alphabet.

Moving back to the island of Great Britain, we first have to stop by on the Isle of Man. There, Manx is spoken by a small number of speakers. It is a critically endangered language, having gone extinct already once in the 1970s. Now there are at least 20 native speakers with some thousand second-language speakers.

On the Isle of Man, Manx is used in some cultural events and now, after long revival efforts, even taught at schools. The well-recorded nature of the language and the similarities in pronunciation to Irish and Gaelic made it easier to revive Manx.

Back to the island of Great Britain, besides Gaelic, there are two more Celtic languages left on the island. One of them, Cornish, has already been a matter of discussion in our part about its sister language, Breton. It is still important to note that in 2010 it has been reclassified as critically endangered and no longer extinct. This was because language revival efforts started in the early 20th century bore fruit.

The only Celtic language remaining on our list is Welsh. Mainly spoken in Wales and some parts of England, it is notably also spoken in YWladfa in Argentina's Chubut Province.

These settlements in Argentina stem from a fear of the Welsh-speaking population that their language might get extinct in Wales itself. Therefore, numbers of Welsh speakers started to immigrate to Argentina, where they started to build their own Welsh-speaking community. Through that the Patagonian Welsh dialect was formed, which, even though its geographic distance, doesn't diverge from Welsh all too much.

Since 2011, Welsh has had an official status in Wales and in general faces a similar situation as Irish. It is supported by the government, but still has to fight against the far stronger influence of English. Nevertheless, it is considered the least endangered language, especially because of the increasing number of children learning Welsh at a young age.

From the Celtic languages we move on to the most spoken language of the British Isles and the world if counted by first and second language speakers combined: English.

There are many ways to describe how widespread English is used, but maybe it's best to take this book as an example, because in the end there is a reason this book is in English and none of the prior named languages. Naturally, writing it in Spanish, German, or French would have opened up the opportunity to speak to a wide audience, but in no case could it be understood by as many people as in English.

Speaking in numbers, it is widely accepted that English is spoken by around 380 million native speakers and an additional billion second language speakers. It can be found as an official or administrative language on every continent and is understood by the majority of the population in quite a few of them. In all, it is an official language in 57 countries and several sovereign entities, many of which are governed under the name of the British crown, and it is used as a working language in countries all around the world.

It is one of the main working languages of many international organisations, including the UNO, EU, NATO, and BRICS. Additionally, it is widely in use in business and politics around the world, even in non-English-speaking countries, and is generally used to communicate with people from a different language background. This widespread usage is primarily due to the global reach of the former British Empire and the United States.

English is the most spoken Germanic language and retains grammar, phonology, and its most used vocabulary from this Germanic origin. Even though this English has been heavily influenced by French and Latin. Loanwords from these two combined make up around 60% of English vocabulary.

Through its development on an island, the development of English differs from that of continental Germanic languages such as German and Dutch. As mentioned earlier, the Frisian languages are the closest relatives of English. Through its divergent evolution, it is not mutually intelligible with any of the continental European languages.

Influence from a variety of places and languages, partly reasoned in the global reach of the language and the influence of those languages on it, which's cultures it influenced, gave English a unique pronunciation and vocabulary.

Some features of the language have become easier through this scheme, while others have become more complex and harder to understand from afar. The former point includes part of its grammar, for example, grammatical gender and articles, while the latter is more focused on pronunciation.

Parts of its grammar can also be seen as tricky or hard, but this is often based on a common misconception that all things in a language have to align with one common norm to be logical. For example, tenses, where verbs are often classified as regular or irregular, which from a language learner's perspective makes only sense. So-called regular verbs follow a common pattern with the ending -ed being added in the past tense. The other irregular verbs do not follow this common pattern, but I would argue that they are still logical. They may not need an -ed at the end, but they still follow some logical pattern; it's just that not all of them follow the same. For example, the verbs put, hit, slit, let, knit, and several more all follow the pattern of not changing; the same can be said about read, which only changes its pronunciation. Other words like send and lend follow the pattern of changing from a soft consonant to a hard one (dàt). This can also be seen in the past tense for go, which is went. It might not seem logical at first—at least not for a second language speaker—but with more background knowledge, it becomes clear that from the Old English verb wend, the past tense went is only logical. Even words like buy, bring, catch, teach, and seek, think follow a specific pattern. Here all words end with -ught with either an a or an o being added depending on if the base word contained an a or a different erent vowel. This change is called ablaut and is a normal Germanic pattern for strong verbs—which, from my perspective, is a far better term to call them.

As said, there is a series of different groupings among the so-called irregular verbs, which makes them logical if one is to think about them a bit longer. This also works for many difficulties in pronunciation, which largely stem from the different origins of words both within and outside the English language.

So, in all this little excursion inside the English grammar, it tells us that it's often important to think about what connections specific words in any language have to understand them better. This is a lesson important when learning any language, not just one from Western Europe.

Our journey therefore ends here in the English-speaking areas of Great Britain, somewhere in the middle of London, a world city where so many languages can be heard in its streets and restaurants that it gives a flair of places far off from England and Europe as a whole.

Sources

Romance languages:

https://theworld.org/stories/2018/05/16/very-fascinating-language-basque
https://www.britannica.com/topic/Portuguese-language
https://cvc.cervantes.es/lengua/anuario/anuario_23/el_espanol_en_el_mundo_anuario_instituto_cervantes_2023.pdf
https://www.babbel.com/en/magazine/how-many-people-speak-catalan
https://blog.rosettastone.com/how-many-people-speak-french-a-full-breakdown-by-country/
https://www.britannica.com/topic/Occitan-language
https://www.britannica.com/topic/Breton-language
https://www.britannica.com/topic/Franco-Provencal-dialect
https://www.britannica.com/topic/Corsican-language
https://www.britannica.com/topic/Sardinian-language#ref603612
https://www.ethnologue.com/language/scn/
https://irpiniastories.com/2020/06/14/the-languages-of-campania/
https://www.babbel.com/en/magazine/how-many-people-speak-italian-where-spoken
https://www.az.com.na/
https://onlinelibrary.wiley.com/doi/10.1111/josl.12283
https://www.cambridge.org/core/books/abs/cambridge-history-of-the-english-language/lexis-and-semantics/04508EFF316B85D7F5ED1AD2DAC902E8

Germanic languages

https://www.britannica.com/topic/Germanic-languages

https://www.britannica.com/topic/Celtic-languages

https://journal.fi/scf/article/view/60553/38319

www.ingramcontent.com/pod-product-compliance
Lightning Source LLC
Chambersburg PA
CBHW070944220526
45469CB00007B/2509